ISBN 978-1-334-33989-9
PIBN 10571952

This book is a reproduction of an important historical work. Forgotten Books uses state-of-the-art technology to digitally reconstruct the work, preserving the original format whilst repairing imperfections present in the aged copy. In rare cases, an imperfection in the original, such as a blemish or missing page, may be replicated in our edition. We do, however, repair the vast majority of imperfections successfully; any imperfections that remain are intentionally left to preserve the state of such historical works.

1 MONTH OF FREE READING

at

www.ForgottenBooks.com

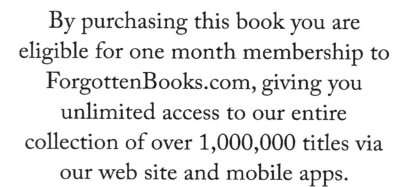

By purchasing this book you are eligible for one month membership to ForgottenBooks.com, giving you unlimited access to our entire collection of over 1,000,000 titles via our web site and mobile apps.

To claim your free month visit:
www.forgottenbooks.com/free571952

Historic, Archive Document

Do not assume content reflects current scientific knowledge, policies, or practices.

United States Department of Agriculture,

BUREAU OF CHEMISTRY—Circular No.

H. W. WILEY, Chief of Bureau.

CHANGES IN OFFICIAL METHODS OF ANALYSIS[a] AND ADDITIONS THERETO, 1899 TO 1905.

INTRODUCTION.

Since the adoption of the official methods of the Association of Official Agricultural Chemists as issued in 1899 the work of the association has been greatly extended in scope and has also become greatly specialized; in other words, the growth of the work has been both extensive and intensive, both along general lines and in the matter of detail. When, therefore, an effort was made to revise the methods by merely incorporating the changes and additions formally authorized by the association in convention, it was found that such a revision would be very inadequate, if not absolutely misleading. The following points are mentioned as illustrative of the character of the changes which should be authorized:

1. In the progress of the work along various lines all under different referees, said referees changing from year to year, new methods have been adopted provisionally which conflict with those under other captions. For example, the very extensive changes made in the methods of sugar analysis have not yet been incorporated under the food adulteration subjects including such determinations (i. e., saccharine products, wines, fruits and fruit products, etc.).

2. Resulting from the same process as that indicated in paragraph 1, there is much unnecessary repetition and overlapping of the methods, especially in the case of the food methods, ash and moisture determinations, etc. This condition calls for the rearrangement of the methods and the grouping together of the general methods to which reference may be made from the specific heads, while under the latter the slight modifications necessary to adapt the general method to the special subject may be given. The following scheme of rearrangement is submitted for consideration:

ARRANGEMENT OF PROPOSED REVISION OF BULLETINS 46 AND 65.[b]

PART I.

Fertilizers.	Waters.
Soils.	Insecticides.
Ash.	Tanning materials.

---

[a] Methods adopted by the Association of Official Agricultural Chemists, November 11, 12, and 14, 1898, and issued as Bulletin No. 46, Bureau of Chemistry.

[b] Provisional methods for the analysis of foods adopted by the Association of Official Agricultural Chemists, November 14, 15, and 16, 1901.

PART II OR PARTS II AND III.

General methods of analysis of foods and feeding stuffs:

Moisture.
Ash.
Fat.
Total nitrogen.
Albuminoid nitrogen.
Amido nitrogen.
Reducing sugars.
Sucrose.
Starch.
Pentosans.
Galactan.
Crude fiber.
Cattle foods.
Cereal foods.
Infants' and invalids' foods.
Vegetables.
Saccharine products.
Sugar-house methods.

General methods of analysis of foods and feeding stuffs—Continued.

Fruit and fruit products.
Wines.
Beers.
Distilled liquors.
Vinegars.
Meat and meat products.
Edible oils and fats.
Dairy products.
Cocoa and its products.
Tea.
Coffee.
Spices.
Condiments other than spices.
Baking powder and baking-powder chemicals.
Food preservatives.
Coloring matter.
Drugs.

3. Some methods remain on the records as provisional which by implication (for example, by the adoption of the methods in modified form) have been nullified, and these should be omitted from the revised methods.

4. The methods should be reworded to insure clarity and uniformity of expression. As the methods and recommendations making changes in them are the work of different referees, many differences in form of expression are found in adopted methods.

As the secretary of the association did not have the authority to make the changes deemed advisable, Circulars 28,[a] 29,[a] and 30 are issued giving a compilation of the authorized changes in and additions to the methods, and all members of the association are urged to consider the present status of the methods and the proposed revision, in order that the convention of 1906 may be ready to act decisively and promptly in regard to this very important matter, the improvement and unification of methods of analysis being the chief object of the association.

H. W. Wiley,
Chief, Bureau of Chemistry,
Secretary, Association of Official Agricultural Chemists.

Approved:

James Wilson,
Secretary of Agriculture.

Washington, D. C., *May 10, 1906.*

---

[a] No. 28, Provisional methods for the determination of food preservatives as authorized by the association, 1905; No. 29, Changes in provisional methods for the analysis of foods and additions thereto, from 1902 to 1905.

The secretary was instructed to place in the next revised edition of the methods the latest revision of atomic weights of the American Chemical Society.

[Note by the Editor.—In making this insertion it is to be noted that all calculations occurring in the methods depending upon atomic weights should be corrected throughout to agree with the new table of atomic weights as adopted.]

I. METHODS FOR THE ANALYSIS OF FERTILIZERS.

[All page references are to Bul. 46.]

Seventeenth Convention, 1900, Bul. 62.

Page 13, under "(b) Optional volumetric method, (1) Preparation of reagents," omit (b) and (c).

Page 14, under "(2) Total phosphoric acid, (b) Determination," add (b_2) and (b_3):

(b_2) Proceed as directed in (b_1), with this exception: Heat in water bath at 45° to 50°, add the molybdate solution, and allow to remain in the bath, with occasional stirring, for thirty minutes.

(b_3) Proceed as in (b_1) to the point where the solution is ready to be placed in the water bath. Then cool solution to room temperature, add molybdate solution at the rate of 75 cc for each decigram of phosphoric acid present, place flask containing solution in shaking apparatus and shake for thirty minutes at room temperature, filter at once, wash, and titrate as in preceding method.

Eighteenth Convention, 1901, Bul. 67.

Page 14, under "4. Determination of nitrogen, (a) Standard hydrochloric acid," the secretary is directed "to correct evident error."

[Note by the Editor.—From correspondence with the chairman of the committee on recommendations who offered this motion it appears that the following changes are desired to make the details of manipulation more definite:

On page 15, first line, for "Allow to cool somewhat, and then filter through asbestos," read, "Allow to cool somewhat, and then filter in a Gooch crucible through asbestos."

For the first sentence of the second paragraph, "Put the capsule and precipitate aside," etc., read as follows: "Put the vessel containing the precipitate aside, remove the filtrate from the receiver and return once through the Gooch crucible so as to obtain it quite clear, and set aside to recover excess of silver."]

(3)

Page 21, under "4. Determination of nitrogen," print the neutral permanganate method [for the determination of available organic nitrogen] as a provisional method:

(k) NEUTRAL PERMANGANATE METHOD (PROVISIONAL).

Into a 400 cc Griffin beaker, low form, weigh an amount of sample containing approximately 0.075 gram of nitrogen (samples containing material that has been treated with acid should be washed on a 9 cm S. S. No. 595 filter to 200 cc and transferred, filter and all, to beaker); digest this with 125 cc of permanganate of potash solution (16 grams of pure $KMnO_4$ to 1,000 cc of water) in a steam or hot-water bath for thirty minutes. Have the beaker let down well into the steam or hot water and keep closed with a cover glass, stirring twice at intervals of ten minutes with a glass rod. At the expiration of the time remove from bath, add 100 cc of cold water and filter through a heavy 15 cm folded filter. Wash with cold water, small quantities at a time, till total filtrate amounts to 400 cc. Dry and determine nitrogen in residue by Kjeldahl method.

Nineteenth Convention, 1902, Bul. 73.

It was earnestly urged by the referee that in reporting analyses of complete fertilizers the amounts of nitrogen existing in the form of nitrates, ammonia salts, and organic nitrogen be specifically stated.

Twentieth Convention, 1903, Bul. 81, Cir. 13.

Page 21, under "4· Determination of nitrogen," change the first two lines of the provisional neutral permanganate method (k) to read as follows:

Into a 300 cc low form Griffin beaker weigh 2 grams of the sample, if from a mixed fertilizer; if from concentrated material use a quantity containing approximately 0.075 gram of nitrogen.

Twenty-first Convention, 1904, Bul. 90, Cir. 20.

Page 21, under "4. Determination of nitrogen," print the alkaline permanganate method for the determination of available organic nitrogen:

(l) ALKALINE PERMANGANATE METHOD (PROVISIONAL).

Weigh out an amount of sample containing 0.045 gram of organic nitrogen and transfer to a 600 cc distillation flask. After connecting with condenser, to which the receiver containing standard acid has been attached, digest with 100 cc of alkaline permanganate solution (16 grams of $KMnO_4$ and 150 grams of sodium hydrate, dissolved in water and made to bulk of 1 liter) for thirty minutes below the boiling point. Then boil until 85 cc of distillate is obtained. If the material shows a tendency to adhere to the sides of the flask, an occasional gentle rotation is necessary during distillation.

II. METHODS FOR THE ANALYSIS OF FOODS.[a]

Seventeenth Convention, 1900, Bul. 62.

The terms "phloroglucol," "diresorcol," and "furfural" were adopted.

Page 25, under "(e) Provisional method for the determination of pentosans by means of phloroglucin," alter the fifth line to read as

[a] See also Cir. 7, Rev., 1902, for amended methods under this section.

follows: "regulated so as to distill 30 cc in about 10 minutes, the distillate passing through a small filter paper."

In the sixth line, after "dilute acid," insert the following: "added in such a manner as to wash down the particles adhering to the sides of the flask."

In the seventh line, after "filter paper," insert the following: "or until the distillate amounts to 360 cc."

In the eighth line change the words "free from diresorcol" to "purified if necessary."

Page 26, to the paragraph entitled "Qualitative test of the purity of phloroglucol," add the method of Fraps for purifying phloroglucol, as follows:

About 300 cc of hydrochloric acid, 1.06 sp. gr., are heated in a beaker and 11 grams commercial phloroglucol added in small quantities at a time, stirring constantly until it has almost entirely dissolved. Some impurities may resist solution, but it is unnecessary to dissolve them. The hot solution is poured into a sufficient quantity of the same hydrochloric acid (cold) to make the volume 1,500 cc. It is allowed to stand at least over night—better several days—to allow the diresorcol to crystallize out and filtered immediately before using. The solution may turn yellow, but this does not interfere with its usefulness. In using it the volume containing the required amount is added to the distillate from the pentosan.

Page 26, "(f) Method for estimating galactan," in line fourteen, after "ammonia," add: "The filter paper and contents are then washed several times with hot water by decantation, the washings being passed through a filter paper, to which the material is finally transferred and thoroughly washed."

Eighteenth Convention, 1901, Bul. 67.

Page 25, "(d) Diastase method for starch," fifth line, instead of "from 20 to 40 cc" read "20 cc"; digest for a definite length of time and without examining with the microscope boil again (microscopic examination to be made later). Sentence as corrected reads as follows:

* * * add 20 cc of malt extract and maintain at this temperature for an hour. Heat again to boiling for a few minutes, cool to 55° C., add 20 cc of malt extract, and maintain at this temperature until a microscopic examination of the residue with iodin reveals no starch.

Line 10, instead of "nearly neutralize while hot with sodium carbonate", read "cool, and neutralize with sodium hydrate or nearly neutralize with sodium carbonate".[a]

Preparation of malt extract, first line, for "overnight" read "two or three hours."

Under "(e) Provisional method for the determination of pentosans by means of phloroglucin," instead of "Three grams of the material are placed in a flask," read "A quantity of the material, chosen so that

---

[a] It appears that this should have been reported as follows: "cool and nearly neutralize with sodium hydrate."

the weight of the phloroglucid obtained shall not exceed 0.300 gram, is placed in a flask".

After the word "added," in the third sentence, beginning "The 30 cc driven over", insert the following: "by means of a separatory funnel and".

Nineteenth Convention, 1902, Bul. 73.

Page 25, under "(e) Provisional method for the determination of pentosans by means of phloroglucin," thirteenth line, "The solution", etc., to end, substitute the following:

The solution is made up to 400 cc with 12 per cent hydrochloric acid, and allowed to stand overnight. The amorphous black precipitate is filtered into a tared gooch through an asbestos felt, washed carefully with 150 cc of water in such a way that the water is not entirely removed from the crucible until the very last, then dried to constant weight by heating four hours at 100° C., cooled, and weighed in a weighing bottle, the increase in weight being reckoned as phloroglucid. To calculate the furfural from the phloroglucid, use the following formulæ:

For weight of phloroglucid (a) under 0.03 grams:

$$\text{Furfural} = (a + 0.0052) \times 0.5170.$$
$$\text{Pentoses} = (a + 0.0052) \times 1.0170.$$
$$\text{Pentosans} = (a + 0.0052) \times 0.8949.$$

For weights of phloroglucid (a) from 0.03 grams to 0.300 grams Kröber's table a is recommended. If this is not available use the formulæ:

$$\text{Furfural} = (a + 0.0052) \times 0.5185.$$
$$\text{Pentoses} = (a + 0.0052) \times 1.0075.$$
$$\text{Pentosa} = ns(a + 0.0052) \times 0.8866.$$

For weights of phloroglucid (a) over 0.300 grams.

$$\text{Furfural} = (a + 0.0052) \times 0.5180.$$
$$\text{Pentoses} = (a + 0.0052) \times 1.0026.$$
$$\text{Pentosans} = (a + 0.0052) \times 0.8824.$$

a J. Landw., 1900 48: 375; Tollens, Zts. angew. Chem., 1902, 15: 477, 508; Zts. physiol. Chem., 1902, 36: 239.

Kröber's factors were adopted (see Journal für Landwirtschaft, 1900, 48:375–384).

Second paragraph, second line, strike out the words "from three to"; third line, insert "in a weighing bottle" after the word "weighed."

Page 26, under "(f) Method for estimating galactan", twelfth line, strike out the word "gently", and for "on a water bath for fifteen minutes" read "on a water bath at 80° C. for fifteen minutes with constant stirring".

Line 14, after the word "bath" insert "avoiding unnecessary heating, which causes decomposition".

Line 19, for "a short time" substitute "three hours".

ERRATA.

Page 24, under "6. Determination of albuminoid nitrogen by Stutzer's method, (a) Preparation of reagent", second line, for "25 cc" read "2.5 cc".

Page 26, top of page, in formulæ I to VI, place minus sign between word "furfural" and factor.

III. METHODS FOR THE DETERMINATION OF SOLUBLE CARBOHYDRATES IN AGRICULTURAL PRODUCTS, ETC.

Eighteenth Convention, 1901, Bul. 67.

Sugar work divided into three parts: Optical methods; chemical methods; special analytical methods.

Twentieth Convention, 1903, Bul. 81, Cir. 13.

An associate referee on molasses was appointed.

Page 27, to the heading "1. Determination of water" add "and density"; after (a) insert "Water"; after "(b) Areometric methods" add "for density and indirect estimation of water".

Page 28, preceding table, insert the following:

This method is accurate with liquids containing only soluble carbohydrates. With syrups, molasses, and liquids containing mineral matter the results are only roughly approximate.

Page 38, insert the following method for ascertaining the amount of copper reduced, as a provisional method:

(f) METHOD BY DIRECT WEIGHING OF THE CUPROUS OXID.—Prepare a Gooch crucible for the filtration by loading it with a felt of asbestos one-fourth inch thick. First thoroughly wash the asbestos with water to remove small particles, then follow successively 10 cc alcohol and 10 cc of ether, and dry the crucible and contents thirty minutes in a water oven at 100° C.

Collect the precipitated suboxid of copper in the felt as usual, thoroughly wash it with hot water, then with 10 cc of alcohol, and finally with 10 cc of ether. Dry the precipitate thirty minutes in a water oven at 100° C., cool and weigh it. The weight of cuprous oxid multiplied by 0.888 gives the weight of metallic copper.

Page 39, Optical methods for the determination of sucrose, omit Clerget's method (page 39) and Creydts' method (page 40), and substitute for them the German official methods, as follows:

Determination of sucrose and raffinose.—This method is not applicable in the presence of optically active bodies other than sucrose and raffinose. Percentages of raffinose less than 0.33 can not be determined with certainty by this method.

Dissolve the normal weight of the material in water, clarify as usual, and dilute to 100 cc. Filter and polarize the filtrate at 20° C. Record the polariscope reading as the direct reading or polarization before inversion.

Take 50 cc of the clarified solution freed from lead, add 25 cc of water in a 100-cc flask, and add little by little, while rotating the flask, 5 cc of hydrochloric acid containing 38.8 per cent of the acid; and heat the contents of the flask, after mixing, on a water bath heated to 70° C. The temperature of the solution in the flask should reach 67° to 70° in two and one-half to three minutes. Maintain a temperature of as nearly 69° as possible during seven to seven and one-half minutes, making a total time of heating of ten minutes. Remove the flask and cool the contents rapidly to 20° C., and dilute the solution to 100 cc. Polarize this solution in a tube provided with a lateral branch and a water jacket passing a current of water around the tube to maintain a temperature of 20° C.

Multiply the invert reading, made at 20° C., by 2. If a preliminary calculation using the formula,

0.7538 × (algebraic difference of the direct and invert readings)=per cent sucrose, gives a percentage which is more than 1 per cent higher than the direct reading, raffinose is probably present, and the following formulæ by Creydt should be used:

P = the direct reading.
I = the invert reading.
S = the percentage of sucrose.
R = the percentage of anhydrous raffinose.

$$S = \frac{0.5188P - I}{0.8454}; \quad R = \frac{P-S}{1.85}.$$

Determination of sucrose in the absence of raffinose.—The inversion and observation should be made as described above and the following formula used:

When the polarizations are made at 20° C.:

$$S = \frac{100(P-I)}{142.66 - \frac{20^\circ}{2}} = 0.7538 \times (P-I).$$

Should the temperature (t) vary from 20°, which is permissible within narrow limits, in the absence of raffinose, use the following formula:

$$S = \frac{100\,(P-I)}{142.66 - \frac{t}{2}}.$$

ALTERNATE OFFICIAL METHOD FOR INVERSION WITHOUT THE APPLICATION OF HEAT.

Take 50 cc of the clarified solution freed from lead, add 5 cc of hydrochloric acid containing 38.8 per cent of acid, set aside during a period of twenty-four hours at a temperature not below 20° C.; or if the temperature be above 25° C. set aside for ten hours. Make up to 100 cc at 20° C. and polarize.

Page 43, add the following provisional methods for direct analysis of sugar beets:

(a) Scheibler's alcoholic method.—In the direct analysis of the beet with the Soxhlet-Sickel apparatus by Scheibler's method, proceed as follows for the extraction of the sucrose:

Place a plug of absorbent cotton in the bottom of the tube, then introduce 26.048 grams of the pulped beet, or 2×16.29 grams, according to the polariscope in use, pressing the pulp lightly with a ròd. Very small fragments of the beet may be used instead of pulp. Connect the extractor with the reflux condenser. Place 75 cc of 95 per cent alcohol in the flask and connect with the extractor; heat the flask in the water bath and continue the extraction from half an hour to two hours or more, according to the state of division of the sample. Use somewhat weaker alcohol if only 16.29 grams of pulp be taken. Cool and remove the flask, substitute a second containing 75 cc of 75 to 80 per cent alcohol, and continue the extraction to ascertain whether the first extraction were complete.

Fill the first flask to the 100-cc mark, after treating the sample with two or three drops of subacetate of lead solution. Mix the contents of the flask, filter, and polarize. Having extracted the normal weight of pulp, the polariscopic reading is the per cent of sucrose in the sample.

The extract in the second flask should also be polarized as a check upon the extraction.

Great care is essential in the polarization of alcoholic solutions. The least quantity of subacetate of lead that will clarify the solution should be used. The solution must be protected from evaporation during the filtration by a cover glass. Avoid irregularities in the temperature of the solution in the observation tube, due to the warmth of the hands; since the density of the solution in different parts of the tube will vary under such conditions, striæ will form, rendering an accurate reading impossible.

The Scheibler method, as above described, differs from the original only in a few minor details, especially in the arrangement of the extraction apparatus. The Soxhlet extraction apparatus is much more effective than Scheibler's original instrument.

(b) *Pellet's aqueous method—Hot digestion.*—Any good rasp may be used in the preparation of the pulp for this method. The special flasks with enlarged necks are convenient for use in this method. Transfer 26.048 grams of the pulp to the flask, using a little water to wash the weighing capsule and funnel; or, for the Laurent, employ 32.58 grams of pulp, i. e., 2 × normal weight. The flasks are graduated to contain 201.35 cc for the Schmidt and Haensch and 201.7 cc for the Laurent polariscopes, in order to compensate for the volume of the mare and the lead precipitate. Add 5 to 10 cc subacetate of lead solution of 54.3° Brix f r the clarification. Approximately 6 to 7 cc are required per 26 grams of beet pulp. This reagent should be run into the flask in advance of the beet pulp. Add a few drops of ether to beat down the foam, then sufficient water to increase the volume of the solution to about 190 cc. Heat to 80° C. in a water bath and maintain this temperature about thirty minutes, occasionally giving the flask a circular movement to facilitate the escape of the air from the pulp. Increase the volume of the solution from time to time during the heating, so that when the operation is completed only a few drops of water will be required to complete the volume of the solution to the mark. After approximately thirty minutes' heating, cool the flask and contents and add strong acetic acid to the solution to acidity, dilute to the graduation, mix, and filter. The state of division of the pulp will govern the time of heating. In polarizing the filtrate, use a 400-mm observation tube, thus directly obtaining the per cent sucrose in the beet with the Schmidt and Haensch polariscope, or double this percentage if the Laurent instrument be used.

Add the methods selected by the international committee for unifying methods of sugar analysis, as *provisional methods* of this association. These methods are as follows:

1. In general, all sugar tests shall be made at 20° C.

2. The graduation of the saccharimeter shall be made at 20° C. Twenty-six grams of pure sugar, dissolved in water, and the volume made up to 100 metric cc, or during the period of transition 26.048 grams of pure sugar in 100 Mohr cc, all weighings to be made in air with brass weights, the completion of the volume and the polarizations to be made at 20° on an instrument graduated at 20° should give an indication of 100 on the scale of the saccharimeter. For countries where temperatures are usually higher than 20° C., it is permissible that saccharimeters be graduated at 30°, or any other suitable temperature, under the conditions specified above, providing that the analysis of the sugar be made at the same temperature—that is, that the volume be completed and the polarizations made at the temperature specified.

3. Preparation of pure sugar: Purest commercial sugar is to be further purified in the following manner: A hot saturated aqueous solution is prepared and the sugar precipitated with absolute ethyl alcohol; the sugar is carefully spun in a small centrifugal machine and washed in the latter with absolute alcohol. The sugar thus obtained is redissolved in water, the saturated solution again precipitated with alcohol and washed as above. The product of the second crop of crystals is dried between blotting paper and preserved in glass vessels for use. The moisture still contained in the sugar is determined and taken into account when weighing the sugar which is to be used.

The committee further decided that central stations shall be designated in each country which are to be charged with the preparation and distribution of chemically pure sugar. Wherever this arrangement is not feasible, quartz plates, the values of which have been determined by means of chemically pure sugar, shall serve for the control of the saccharimeters. The committee further decided that the above control of quartz plates by means of chemically pure sugar should, as a rule, apply only to the central stations which are to test the correctness of saccharimeters; for those who execute commercial analyses, the repeated control of the instruments is to be accomplished, now as before, by quartz plates.

4. In effecting the polarization of substances containing sugar, half-shade instruments, or triple field, only are to be employed.

5. During the observation the apparatus must be in a fixed position and so far removed from the source of light that the polarizing nicol is not warmed.

6. Sources of light may be gas, triple burner with metallic cylinder, lens and reflector; gas lamps with Auer (Welsbach) burner; electric lamp; petroleum duplex lamp; sodium light. Several readings are to be made and the mean thereof taken, but any one reading must not be neglected.

7. In making a polarization, the whole normal weight for 100 cc is to be used, or a multiple thereof for any corresponding volume.

8. As clarifying and decolorizing reagents there may be used: (a) Subacetate of lead (3 parts by weight of acetate of lead, 1 part by weight of oxid of lead, 10 parts by weight of water); (b) alumina cream; (c) concentrated solution of alum. Boneblack and decolorizing agents are to be excluded.

9. After bringing the solution exactly to the mark, at the proper temperature, and after wiping out the neck of the flask with filter paper, all of the well-shaken clarified sugar solution is poured upon a dry rapidly acting filter. The first portions of the filtrate are to be rejected and the rest, which must be perfectly clear, used for polarization.

ERRATA.

Page 27, under "1. Determination of water, (a) By drying, (3) Provisional method for drying molasses with quartz sand," third line from the end of section, for "(See method II (d) under 'ash')," read "(See method (a) (5), page 31, under '2. Ash')."

Page 30, three lines from bottom of page, for "50°" read "50".

Page 39, formula under "(b) Optical methods by inversion, (1) Method of Clerget," should read "100(a—b)" in the numerator instead of "a—b."

Page 40, fourth line from top of the page, under "Creydt's method for determining raffinose and sucrose together," transfer "at 20°" to the end of the sentence.

IV. METHODS FOR THE ANALYSIS OF DAIRY PRODUCTS.

Seventeenth Convention, 1900, Bul. 62.

Page 47, under "(2) Determination, (a) Soluble acids," insert in the third line, after "saponified," the following: "or the saponification may be conducted in a flask connected with a reflux condenser."

Page 48, under "(h) Determination of saponification equivalent (Koettstorfer number), (2) Determination," change the words "Between 1 and 2 grams" to "Approximately 2.5 grams". At the end of the same sentence, insert the following after a semicolon: "or the saponification may be conducted in a flask with a reflux condenser".

Page 50, make such changes in the quantities of sodium thiosulphate as will be consistent with the new atomic weights.

Page 51, under "Optional method of standardizing flasks," insert the following in place of the method there given:

The following formula may be used for calculating the weight of water ($W^{T\circ}$) which a given flask will hold at T° (weighed in air with brass weights at the temperature of the room) from the weight of water ($W^{t\circ}$) (weighed in air with brass weights at the temperature of the room) contained therein at t°:

$$W^{T\circ} = W^{t\circ} \frac{d^{T\circ}}{d^{t\circ}} (1 + \gamma (T - t))$$

$d^T =$ the density of water at T°.
$d^t =$ the density of water at t°.
$\gamma =$ the coefficient of cubical expansion of glass.

(See also "Separation of Milk and Cheese Proteids," page 20 of this circular.)

Page 52, under "(1) Determination of melting point—Wiley's method, (3) Determination," end of the first paragraph, insert as follows: "These disks should be prepared a day, or at least a few hours, before using."

Page 55, change paragraph 3 to read as follows:

3. *Official optional method for determining casein.*—To 10 cc of milk add 50 cc of distilled water at 40°, then add 2 cc of alum solution saturated at 40° or higher. Allow precipitate to settle, transfer to a filter, and wash. The washed precipitate and filter paper are digested as in regular Kjeldahl method, as outlined in the official method, paragraph 1. Calculate casein by multiplying the amount of nitrogen found by 6.25.

Twenty-second Convention, 1905, Bul. 99, Cir. 26.

Page 49, change correction for temperature for index of refraction from 0.000176 to 0.000365. [Resultant corrections should be made.]

ERRATA.

Compare the following: Page 43, section (c) (2), second line, "or with 76° petroleum ether", with page 54, section 2. (b) (1), next to the last line, "ether or petroleum ether boiling at about 45°".

V. METHODS FOR THE ANALYSIS OF FERMENTED AND DIS-TILLED LIQUORS.

No changes have been made in this section except in so far as it has been modified in practice by the provisional methods introduced in the food adulteration work, which methods are given in Bulletin 65, and in Circular 29, giving changes in and additions thereto.

ERRATUM.

Page 64, section (8) (a) fourth line, for "11" read "12".

VI. METHODS FOR THE ANALYSIS OF SOILS.

Seventeenth Convention, 1900, Bul. 62.

Page 71, fourth paragraph, use a 3-mm sieve when the determinations are to be made on 100 grams or more of soil.

Page 74, under "(i) Provisional method for the determination of the more active forms of phosphoric acid in soils," make determinations of phosphoric acid by fifth normal *nitric* acid as well as by fifth normal *hydrochloric* acid.

Eighteenth Convention, 1901, Bul. 67.

Page 71, following "1. Preparation of sample," insert the following:

All samples of soils taken for analysis should be composite and should be composed of representative samples taken from at least five different places in the field sampled, each individual sample to be a column of uniform soil extending through the stratum sampled.

One composite sample should be taken from each important and distinctly different soil stratum to a depth of 40 inches, or 1 meter, including a composite sample from the arable stratum, or plowed soil, usually about 6 inches or 15 cm deep.

If the plow line and the subsoil line coincide, and the subsoil is a fairly uniform stratum to a depth of 40 inches, then only two composite samples need be taken, one of the arable soil and one of the subsoil. But if the subsoil line is lower than the plow line and not below 40 inches, then both strata below the arable soil should be sampled, which would make three strata to be sampled and necessitate the taking of three composite samples from the field—one from the surface or arable soil; one from the subsurface soil, that is, from the stratum between the plow line and the true subsoil line, and one from the true subsoil.

Nineteenth Convention, 1902, Bul. 73.

Page 71, so amend the provisional method for soil sampling (Hopkins) that it will be possible to determine the available potash and phosphoric acid in each stratum sampled to a total depth of 3 or 4 feet.

Page 77, insert the following quantitative method for the determination of the acidity of soils as provisional:

PROVISIONAL QUANTITATIVE METHOD FOR DETERMINING THE ACIDITY OF SOILS (HOPKINS, KNOX, AND PETTIT).

Place 100 grams of the soil in a stout, medium wide-mouthed bottle of about 400-cc capacity; add a sufficient quantity of a 5 per cent sodium chlorid solution to make 250 cc of liquid, including the moisture contained in the soil, but independent of the volume of the soil itself (the moisture content of thoroughly air-dry soils will not cause an appreciable error, if neglected); close tightly with a rubber stopper; place in a shaking machine, and shake for three hours. (Shaking by hand every half hour for about twelve hours will accomplish the same result.) Place the bottle in a suitable centrifuge and whirl till the soil is thrown down sufficiently to allow at least 125 cc of clear liquid to be blown off by means of two glass tubes held in a rubber stopper and arranged about the same as in the ordinary "spritz" bottle, except that the exit tube should be about 2 mm at the end instead of terminating in a fine jet. Instead of using the centrifuge the bottles may be allowed to stand a few hours, or 125 cc of the liquid may be filtered off, care being taken to avoid concentration by evaporation. Exactly 125 cc of the clear liquid (one-half the total volume of liquid) are placed in an Erlenmeyer flask, heated to boiling to expel traces of carbon dioxid and to insure a sharp "end reaction," and then titrated with standard fixed alkali, phenolphthalein being used as the indicator.

Twentieth Convention, 1903, bul. 81, Cir. 13.

Page 72, under "4 (a) Acid digestion of the soil," eleventh line, for "and again evaporate to complete dryness" substitute the following: "filter, wash free of chlorids, and again evaporate to complete dryness as before."

Page 74, under "4 (g) Determination of phosphoric acid," mark the official method (a) and insert the following:

(b) *Optional provisional method:* Proceed as in (a) "until all the phosphoric acid is precipitated," and then finish the determination as follows:

After standing for three hours at a temperature not above 50°, filter on a small filter, wash with water until two fillings of the filter do not greatly diminish the color produced with phenolphthalein by one drop of standard alkali. Place the filter and precipitate in the beaker and dissolve in standard alkali, add a few drops of phenolphthalein solution and titrate with standard acid, 1 cc of which equals 0.0005 gram of phosphoric acid (P_2O_5).

Page 74, "4 (h) Provisional method for determining available phosphoric acid," omit the word "available" from the heading.

Page 76, under "10. Determination of humus," tenth line, change the sentence beginning "The supernatant liquid" to read as follows:

The supernatant liquid is filtered, and the filtrate must be perfectly clear and free from turbidity; evaporate an aliquot portion, dry at 100°, and weigh.

[The retiring referee was instructed to make necessary verbal changes in the methods and submit them to the secretary for publication.]

VII. METHODS FOR THE ANALYSIS OF ASHES.

Eighteenth Convention, 1901, Bul. 67.

Page 77, name of section is changed to "VII. Methods for the Determination of Inorganic Plant Constituents."

Page 79, insert as provisional the modified nitric acid method for the determination of sulphur in plants:

MODIFIED NITRIC-ACID METHOD FOR THE DETERMINATION OF SULPHUR IN PLANTS.

Place 5 grams of material in a 3½-inch porcelain evaporating dish, add 20 cc of nitric acid (conc.), and heat the mixture cautiously on a water bath until all danger of overflowing is passed. Partly evaporate, add 10 cc of a 5 per cent solution of potassium nitrate, evaporate the mixture to complete dryness, and ignite, at first gently and then under a blast lamp, until the residue is white. Then dissolve in hydrochloric acid, evaporate to dryness, and heat for some time in an air bath to render silica insoluble. Take up the residue in water, with the addition of a little acid, filter, and precipitate the sulphuric acid with barium sulphate, etc., in the usual way.

[This method was dropped in 1905.]

Page 79, section "7· Determination of chlorin" is to be omitted until a more exact method can be offered (refers to chlorin in the plant, according to new name of section, not to chlorin in ash).

Page 79, add the determination of potash by ignition of the plant substance with sulphuric acid, as described in the determination of potash in fertilizers [p. 22, (3) Determination, (a) In mixed fertilizers,] as an alternative method.

The entire methods for the analysis of ashes as rewritten by the referee were adopted, as follows:

VII.—METHODS FOR THE DETERMINATION OF INORGANIC PLANT CONSTITUENTS.

1. PREPARATION OF SAMPLE.

The material must be thoroughly cleaned from all foreign matter, especially from adhering soil. It is to be ground and preserved in carefully stopped bottles.

2. DETERMINATION OF CARBON-FREE ASH.

(a) *Preparation of calcium acetate.*

Dissolve 20 grams c. p. calcium carbonate in c. p. acetic acid, and dilute to a liter. Evaporate 20 cc in a platinum dish, ignite gently, then strongly, to constant weight. The dish must be weighed quickly. This gives the calcium oxid in 20 cc.

An alternative method is to dissolve marble in hydrochloric acid, evaporate, and dry to render silica insoluble, dissolve with water and a little acid, and precipitate iron and aluminum in the usual way. The calcium is then precipitated with ammonia and ammonium oxalate in hot solution, the precipitate washed well, dried, ignited, and weighed. It is then dissolved, and diluted so that 100 cc contains 1.1 grams calcium oxid.

It is best to test the purity of this reagent by making blank determinations with it.

(b) *Preparation of ash.*

Moisten 10 to 20 grams substance with 40 cc calcium acetate, dry on a water bath, and ignite, gently at first, then more vigorously. The quantity of calcium acetate used should be sufficient to prevent fusion of the ash. Some form of apparatus must be used to prevent volatilization, either Shuttleworth's [a] or Tucker's [b] or an ordinary platinum dish may be used, fitted with a cover, like that described by Wislicenus.[c] The weight of the ash must be corrected for lime, carbon dioxid, and carbon.

[a] Exp. Sta. Rec., **11**: 304.
[b] Ibid., 506.
[c] Zts. anal. Chem., 1901, **40**: 441.

(c) *Determination of carbon dioxid.*

Using the ash prepared in (b), liberate the carbon dioxid with hydrochloric acid in any of the usual forms of apparatus, determining the carbon dioxid evolved either by increase of weight of potash bulbs or loss of weight of the apparatus. The former method is preferred.

(d) *Determination of carbon, sand, and silica.*

The residue from the carbon dioxid determination is transferred to a beaker or evaporating dish, evaporated to dryness, and thoroughly dried and pulverized to render silica insoluble. The dry residue, etc. [Same as p. 78, paragraph 2, to the end.]

Subtract carbon, carbon dioxid, and calcium oxid added in the form of calcium acetate from the ash, and calculate results as carbon-free ash.

3. DETERMINATION OF MANGANESE, CALCIUM, AND MAGNESIUM.

As on page 78 of the "Methods," unchanged, save that the word "ash" is inserted after "grams," line 2; add "The quantity of calcium found must be corrected for the calcium added."

4. DETERMINATION OF PHOSPHORIC ACID.

(a) An aliquot portion of the hydrochloric-acid solution, corresponding to 0.2 to 1 gram ash, is to be used for the determination by any of the methods described for total phosphoric acid in fertilizers.

(b) The determination can also be made in the plant substance as in (a_2), page 12 of the "Methods," for phosphoric acid, using sufficient material to give from 0.2 to 1 gram ash in the aliquot portion of the solution taken for the phosphoric acid determination.

5. DETERMINATION OF ALKALIES.

(a) An aliquot portion of the hydrochloric-acid solution (see 2) corresponding to 0.5 to 1 gram of ash is evaporated to dryness. Redissolve the residue in about 50 cc water, add milk of lime or barium-hydroxid solution, which must be perfectly free from alkalies, until no further precipitation is produced.

(b) Potash may be determined as directed for potash in organic compounds (b), page 22 of the "Methods," using sufficient plant material to get from 0.5 to 1 gram ash in the aliquot portion of the solution taken for the potash determination.

6. DETERMINATION OF SULPHUR (PROVISIONAL METHOD).

Modified nitric-acid method as given above.

Nineteenth Convention, 1902, Bul. 73.

Insert the method of determining chlorin by ignition with sodium carbonate as a provisional method:

7. DETERMINATION OF CHLORIN IN PLANTS BY IGNITION WITH SODIUM CARBONATE (PROVISIONAL.)

Five grams substance in a platinum dish are impregnated with 20 cc of a 5 per cent solution of sodium carbonate, evaporated to dryness, and ignited as thoroughly as possible. The residue is extracted with hot water, filtered, and washed. It is returned to the platinum dish, ignited to an ash, dissolved in nitric acid, and the chlorin determined by the Volhard method.

Twentieth Convention, 1903, Bul. 81, Cir. 13.

The following method for the determination of sulphates is adopted as provisional:

PROVISIONAL METHOD FOR THE DETERMINATION OF SULPHATES.

Five grams substance are mixed well with 50 cc of a 1 per cent solution of hydrochloric acid, allowed to stand half an hour, filtered, and washed with the dilute acid to 250 cc or more. The liquid is heated to boiling, barium chlorid is added, and the determination completed in the usual way.

The provisional nitric-acid method for sulphur in plants was dropped.

The peroxid method was adopted as provisional for the determination of sulphur in plants:

PROVISIONAL PEROXID METHOD FOR DETERMINATION OF SULPHUR IN PLANTS.

Weigh out 15 grams of sodium peroxid (free from sulphates) and introduce about three-fourths of it at once into a 100 cc nickel crucible; add a little water and boil over a sulphur-free flame until excess of water is completely driven off. Allow to cool until pasty and stir 2 grams of material into it as quickly as possible; heat cautiously until danger of foaming has passed, adding the remainder of the peroxid in small portions from time to time to complete the oxidation. After fusion is complete, allow crucible to cool; dissolve contents in water, transfer to a beaker, and add a slight excess of hydrochloric acid. Heat to boiling, filter if not perfectly clear; dilute to about 400 cc and add 10 cc of a 10 per cent solution of barium chlorid. Let stand over night, filter, and weigh the barium sulphate in the usual manner. Make blank test with the reagents.

VIII. METHODS FOR THE ANALYSIS OF TANNING MATERIALS.

Seventeenth Convention, 1900, Bul. 62.

Page 79, "III. Moisture, (b), 2," for "103°" read "110°".

Page 79, "IV. Total solids," first line, for "100 cc" read "50 cc".

Page 80, "V. Soluble solids," first line, for "double-folded filter" read "double-pleated filter"; second line, for "100 cc" read "50 cc"; fourth line, for "kaolin" read "barytes". After the paragraph insert the following optional method:

Optional method.

(a) To 100 cc of the solution add 10 cc of a solution of lead acetate (4 grams per liter), adding the reagent drop by drop from a burette and stirring meanwhile. Add 10 cc of a solution of acetic acid (36 grams glacial acid per liter), stirring. Throw on double-pleated filter, reject until clear, and evaporate and dry 50 cc

(b) On another portion of 100 cc repeat the foregoing, except that 20 cc of lead acetate solution shall be used.

Residue from (a) shall be multiplied by 1.2, and from (b) by 1.3, to bring back to the original 100 cc. Add to the corrected weight of the residue from (a) the difference between (a) and (b) and calculate the found residue to soluble solids This corrects for dilution and removal by lead.

Page 80, "VI. Non-tannins," for lines 6–9 substitute the following:

Weigh the remaining three-quarters, which must contain between 12 and 13 grams of dry hide, add to 200 cc of the solution, and shake 10 minutes. Throw on funnel with cotton plug in stem, return until clear, evaporate 50 cc, and dry.

Page 80, "VIII. Testing hide powder, (a)," fourth line, for "100 cc" read "50 cc" and for "10 mg" read "5 mg"; "(b)," second line, for "100 cc" read "50 cc".

Add paragraph (c), as follows:

(c) Any analysis made with a powder which does not fulfill the conditions of the preceding paragraphs shall not be reported as by this method.

Eighteenth Convention, 1901, Bul. 67.

Page 79, under "II. Quantity of material," first line, substitute "0.35 to 0.45" for "0.8"; second line, substitute "tannin" for "solid"; sixth line, substitute "0.35 to 0.45" for "0.8," and "tannin" for solids".

Under "III. Moisture, (b) 1," substitute "8" for "24"; "(b) 2," substitute "6" for "8"; strike out "to 110°".

Under "IV. Total solids," for "50 cc" read "100 cc".

For paragraph "V. Soluble solids," substitute the following:

Double-pleated filter paper (S. and S., No. 590, 15 cm) shall be used. To 2 grams of kaolin add 75 cc of the tanning solution, stir, let stand fifteen minutes, and decant as much as possible. Add 75 cc more of the solution, pour on filter, keep filter full, reject the first 150 cc of filtrate, evaporate the next 100 cc, and dry. Evaporation during filtration must be guarded against.

The optional method under this head, adopted in 1900, was stricken out.

Under "VI. Nontannins," for the first two lines substitute the following:

Prepare 20 grams of hide powder by digesting twenty-four hours with 500 cc of water and adding 0.6 gram chrome alum in solution, this solution to be added as follows: One-half at the beginning and the other half at least six hours before the end of the digestion. Wash by squeezing through linen, continue the washing until the wash water does not give a precipitate with barium chlorid.

Insert the following provisional method:

To 14 grams of dry chromed hide powder in a shaker glass add 200 cc of the tanning solution; let stand two hours, stirring frequently; shake fifteen minutes; throw on funnel with a cotton plug in the stem, let drain, tamp down the hide powder in the funnel, return the filtrate until clear, and evaporate 100 cc.

Throughout the method substitute 100 for 50 cc to allow for evaporation.

Nineteenth Convention, 1902, Bul. 73.

Page 79, paragraph III, omit moisture determinations in extracts.

Change paragraph "(b)" under "III. Moisture" to read as follows:

Evaporations shall take place under precisely the same conditions as to contact with steam or with a metallic plate; all dryings called for after evaporation shall be done by one of the following methods under precisely the same conditions, so that the different residues of each analysis may occupy the same shelves in the drying oven:

1. For eight hours at the temperature of boiling water in steam oven.
2. For six hours at 100° C. in air bath.
3. For five hours at 100° C. in vacuo.

Page 80, "V. Soluble solids," is to read as follows:

Single-pleated filter paper (S. and S., No. 590, 15 cm) shall be used. To 2 grams kaolin add 75 cc of the tanning solution, stir, let stand fifteen minutes, and decant as much as possible (not on the filter), add 75 cc of the solution, stir, and pour on the filter. Keep filter full, reject the first 150 cc of filtrate, evaporate, and dry next 100 cc. The portion dried for determination shall be perfectly clear, and evaporation during filtration must be guarded against.

Under "VI. Nontannins," use 2 grams kaolin when filtering. Omit paragraph "VIII. Testing hide powder,"and "(b)" under "IX. Testing nontannin filtrate."

The following provisional method for the determination of total acidity in liquors was adopted:

DETERMINATION OF TOTAL ACIDITY IN LIQUORS.

Place 100 cc of the liquor in a 500-cc flask and make up to the mark with water. To 100 cc of diluted liquor in a flask with tube condenser add 2 grams of chemically pure animal charcoal. Heat to boiling temperature with frequent shaking, cool, filter, and titrate an aliquot portion with decinormal alkali.

Twenty-first Convention, 1904, Bul. 80, Cir. 20.

Page 79, under "II. Quantity of material," change the last sentence to read as follows: "In the case of extracts, weigh in a weighing bottle fitted with a ground glass stopper such quantity," etc.

Page 80, the following method for the analysis of barks, woods, etc., was adopted as provisional:

PROVISIONAL METHOD FOR BARKS, WOODS, LEAVES, ETC.

I. *Moisture determination.*

Immediately upon receipt of sample grind or crush as finely as possible 25 grams, and dry for 12 hours at 100° C., reweigh, and calculate difference as H_2O.

II. *Preparation of sample for extraction.*

Barks, woods, leaves, and other tanning materials, such as nut galls, myrobalans, etc., should be dried sufficiently to facilitate grinding and then ground to such a degree of fineness that the sample will pass through a sieve of 14 meshes to the inch (linear).

In case of a coarser residue being left after sieving the portion ground, it should be reground until the entire quantity passes through the sieve. A style of grinder should be used that will make a granular preparation of the desired size, not finer. An excess of dust and very fine particles must be guarded against in order that a rapid and complete extraction be obtained.

III. *Quantity of sample to be used for extraction.*

Such a quantity of the material to be analyzed should be taken as will yield a sufficient volume of liquor having a specific gravity of 1.0025 at 15° C. Should the gravity of the liquor prove greater than the above, dilute to the desired point with water about 80° C. If, however, it should prove impracticable to obtain from spent material the necessary volume of liquor of the gravity specified, the gravity should be taken and the quantity of hide powder used to precipitate the tannin, reduced in proportion.

IV. *Extraction of sample.*

The extraction may be carried out in any form of extractor that will give complete extraction, with a yield of not more than 1 liter of liquor. A temperature of 100° C. should be maintained throughout the entire operation, except in the case of sumac and starch-bearing materials, when a maximum of 80° C. will be found better, and extraction continued until resulting liquor shows no precipitation with gelatin solution, as by the method for testing nontannins. Should the volume of liquor prove insufficient, dilution should be made immediately with water at 80° C. to as near the mark as possible.

In case of pulverized or finely ground material, such as sumac, powdered nut galls, etc., the extraction will be materially assisted by thoroughly mixing the sample with glass sand free from iron and matter precipitable or soluble in hot tannin solution.

V. *Total solids.*

After extraction and final dilution cool slowly to about 20° C., shake thoroughly, allow to stand for one hour, then pipette 100 cc into a tared dish, evaporate, dry, and weigh according to method in Bulletin 46.

VI. *Soluble solids, nontannins and tannin.*

The determination of the above items should be conducted as by the method in Bulletin 46.

NOTES.—In all cases it is best to conduct the analyses from a dry basis—that is, with the material dried at 100° C. If the analysis is made to determine the tanning or commercial value, results should be calculated back to the natural condition of the material. In the case of analyses made for comparison between fresh and spent materials and in controlling tannery or extract manufacturing processes results are more conveniently calculated and recorded on a dry basis.

Attention must be given to the quantity of hide powder used in determining nontannins. See Paragraph III, Quantity of Sample, etc., above.

The method for determination of acid in tannin solutions made provisional in 1902 was continued as such with the following alternative method:

ALTERNATIVE PROVISIONAL METHOD FOR DETERMINATION OF ACID IN TANNIN SOLUTIONS.

To 100 cc of diluted liquor, prepared as in the provisional method, add 2.5 grams chemically pure animal charcoal, stir every fifteen minutes for two hours, filter, and titrate an aliquot portion with decinormal alkali.

Page 80, under "V. Soluble solids," change the last sentence to read as follows: That the funnel shall be covered with a plate during filtration, and all other means be taken to guard against evaporation.

Twenty-second Convention, 1905, Bul. 99, Cir. 26.

Under "VI. Nontannins" (addition of 1901), omit "this solution to be added as follows: One-half at the beginning and the other half at least six hours before the end of the digestion," and substitute therefor, "at one time."

In third sentence, after the word "press," insert "limiting the moisture content of wet chromed hide powder used for analysis to 70 to 75 per cent."

The following method for the analysis of liquors was adopted as provisional:

PROVISIONAL METHOD FOR THE ANALYSIS OF LIQUORS.

X. ANALYSIS OF LIQUORS.

Provisional method.

Liquors must be diluted for analysis, so as to give as nearly as possible 0.7 grams of solids per 100 cc.

Total solids to be determined as in extract analysis.

Soluble solids determination to be made in the same manner as with solutions of extracts.

Nontannin determinations must be made by shaking 200 cc of solution with a proportion of wet chromed hide powder containing 70 to 75 per cent moisture, according to the following table:

Tannin range per 100 cc.	Dry hide per 200 cc solution.
Grams.	Grams.
0.35–0.45	10 –12
.25– .35	7 –10
.15– .25	4¼– 7
.00– .15	0 – 4¼

100 cc must be evaporated in each case, the method of evaporation and drying pursued to be the same as for extract analysis.

Under "III Moisture" (Bul. 46) or "IX. Method of drying," Cir. 8 Revised, insert "4· For 16 hours in the combined evaporator and dryer."

[Motion to provisionally adopt the form of apparatus known as the combined evaporator and drier.]

SEPARATION OF NITROGENOUS BODIES.

MILK AND CHEESE PROTEIDS.

Nineteenth Convention, 1902, Bul. 73.

[Note by the Editor.—Attention is called to the methods for the separation of nitrogenous bodies in milk and cheese, recommended for adoption as provisional by the referee, Mr. L. L. Van Slyke, in 1902. (Bul. 73, pp. 87–98.) These methods were referred to Committee B shortly before adjournment, and no action has been taken on them. (See also page 176 of Bul. 73.)]

Twentieth Convention, 1903, Bul. 81, Cir. 13.

The following method for the separation and determination of casein monolactate and casein dilactate was adopted as a provisional method:

Casein monolactate and casein dilactate.

1. *Determination of casein monolactate in milk.*—Casein monolactate in milk coagulates readily at 40° C. Hence in a milk containing only casein and casein monolactate the monolactate can be separated by heating the milk to about 40° C., filtering the precipitate formed, washing, and determining the nitrogen in the precipitate. Ten grams of milk diluted with 90 cc of water give good results.

2. *Separation of casein monolactate and casein dilactate.*—Casein dilactate coagulates completely at 40° C. and below. In milk containing casein monolactate and dilactate, heat 10 grams of milk diluted with 90 cc of water to 40° C., and these two salts precipitate and are separated from the milk casein by filtration. The washed precipitate is then treated with 100 cc of a 5 per cent solution of sodium chlorid, and the whole heated to 55° C., with frequent agitation for two hours. The process is facilitated somewhat by the presence of pure quartz sand. The casein monolactate goes into solution and is separated from the casein dilactate by filtration and washing.

3. *Separation and determination of casein, casein monolactate, and casein dilactate in milk.*— (a) The total amount of nitrogen precipitated by acid is determined by the official method prescribed for determining casein in milk. (Bul. 46, p. 55.)

(b) Heat 10 grams of milk diluted with 90 cc of water to 40° C. for fifteen or twenty minutes, filter the precipitate formed, and wash with distilled water. Transfer the precipitate to a small Erlenmeyer flask, provided with a stopper, treat with 100 cc of a 5 per cent solution of sodium chlorid and heat at 55° C. for two hours, with frequent agitation. Filter the mixture, wash the remaining precipitate with water, and determine the nitrogen in both the precipitate and filtrate. The nitrogen in the precipitate represents casein dilactate; that in the filtrate, casein monolactate. The sum of these two subtracted from the total nitrogen found by precipitation with acid gives the amount of nitrogen as casein.

Twenty-second Convention, 1905, Bul. 99, Cir. 26.

The following changes were adopted in the proposed provisional methods offered in 1902, as explained above:

The amount of water used in making the extract is increased from 500 cc to 1,000 cc.

After the fat and insoluble nitrogenous bodies have been removed by the absorbent cotton, pass the filtrate through asbestos.

SEPARATION OF VEGETABLE PROTEIDS.

Twenty-first Convention, 1904, Bul. 90, Cir. 20.

(No attempt shall be made, on the results of the following method, to separate the proteids of wheat into individuals, and only such names are to be applied as describe the process of separation, viz, (a) proteids soluble in alcohol; (b) proteids insoluble in alcohol; (c) proteids soluble in dilute salt solution.)

PROVISIONAL METHOD FOR THE SEPARATION OF THE PROTEIDS OF WHEAT.

A. *Total Proteids.*—Determine total nitrogen by the Gunning method. Nitrogen×5.68 = total proteids.

B. *Proteids Soluble in Alcohol.*—1. Weigh out 2 grams of flour into a small flask holding only slightly more than 100 cc, about 110 cc. Add 100 cc of 70 per cent alcohol and shake thoroughly for fifteen minutes. Allow to stand eighteen hours and filter into a Kjeldahl digestion flask. Wash the residue with 100 cc of alcohol. Add to the total filtrate 5 to 10 cc of concentrated sulphuric acid and distil off the excess alcohol, continuing the distillation until fumes appear. Determine nitrogen by the Gunning method:

$$\text{Nitrogen} \times 5.68 = \text{proteids soluble in water.}$$

2. Determine proteids soluble in alcohol by Snyder's polariscopic method. (J. Amer. Chem. Soc., 2 : 263.)

C. *Proteids Insoluble in Alcohol.*—Subtract proteids soluble in alcohol from total proteids to obtain proteids insoluble in alcohol.

D. *Proteids Soluble in Dilute Salt Solution.*—Use a 5 per cent solution of potassium sulphate. Place 4 grams of the flour in a flask such as is used in B 1, and introduce exactly 100 cc of the salt solution. Shake thoroughly and allow to stand, with frequent shakings, for eighteen hours, or, better still, agitate in a shaker for six hours. After standing to settle, filter off 50 cc of the liquid, pouring back the first portions of the filtrate until it filters clear, and determine the nitrogen therein by the Gunning method. The amount multiplied by 2 gives the nitrogen in the original flour. This result multiplied by 5.68 gives the amount of proteids in the flour soluble in a 5 per cent potassium sulphate solution.

Twenty-second Convention, 1905, Bul. 99, Cir. 26.

PROVISIONAL METHOD FOR THE SEPARATION OF NITROGENOUS BODIES IN BARLEY AND MALT.

[Condensed description of the method as given by H. T. Brown, in the "Transactions of the Guinness Research Laboratory," 1903, I (1): 61, for The Estimation in Barley and Malt of the Total Amount of Nitrogenous Substances Soluble in Water.]

In order to have the results on the barley and on the malt comparable, the percentage results obtained on the malt are converted back to the equivalent weight of its original barley. To do this, it is necessary to know the weight of 1,000 kernels of dry barley and of 1,000 kernels of dry malt. This weight of dry barley being expressed by Wb and of dry malt by Wm, any percentage weight of a constituent in the dry malt is converted back to its original barley by multiplying by the factor $\frac{Wm}{Wb}$.

It was found that the relation of the weight of grain to that of the extraction water was a very important factor, owing to the different solubility of globulins in salt solutions of varying strength, and in order to make this constant the following conditions were established:

The weight of barley and of malt taken was such that when the solution was made to its final volume it contained the extract and insoluble portions of amounts corresponding exactly to 20 grams of dry barley in 100 cc. For the extraction 75 per cent of the required amount of water was taken at first, the mixture shaken in a revolving shaker for a definite time, the flask then filled to the required volume, and after thorough mixing allowed to stand for six hours, measuring the time from the beginning of the shaking. It was found that six hours' extraction removed all of the nitrogenous substances soluble in cold water and that any decided increase in the length of time of extraction was liable to introduce errors due to the action of proteolytic enzymes.

In order to obtain, by calculation, the amount of malt, which, in the final volume of the mixture of malt and water, shall contain in 100 cc the exact equivalent of 20 grams of dry barley, the moisture content of both barley and malt must be known; also the relation of the dry weight of 1,000 kernels of barley to that of the resulting malt. Knowing these factors, the following general formula is obtained:

$$\frac{Wm \cdot C \cdot V}{Wb\ (100-Mm)} = X,$$

in which

X = weight of air-dry malt to be taken.
Wm = weight of 1,000 kernels of dry malt.
Wb = weight of 1,000 kernels of dry barley.
Mm = percentage of moisture in air-dry malt.
C = required concentration in terms of grams of dry barley per 100 cc of the final mixture.
V = volume of final mixture in cubic centimeters.

After the extraction is completed the contents of the flask is filtered and definite volumes (100 cc) of the clear filtrate taken. In one of these portions the amount of nitrogen is determined by the Kjeldahl method. This represents the total soluble nitrogenous compounds. The other portion of the filtrate is boiled down to small bulk, in a beaker, again made up to its original volume and filtered. The nitrogen in this filtrate represents the soluble nitrogenous compounds not coagulable by boiling, and the difference between this and the total nitrogen is the nitrogen representing the nitrogenous substances coagulable by boiling.

SEPARATION OF MEAT PROTEIDS.

Twenty-first Convention, 1904, Bul. 90, Cir. 20.

PROVISIONAL METHODS FOR THE DETERMINATION OF AMMONIA, CHLORIN, AND PHOSPHORIC ACID IN MEAT EXTRACTS.

Ammonia.—By the magnesium oxid method, as described in Bulletin 46, page 21.

Acidity.—Titrate with standard alkali solution, using litmus paper as an indicator. The solution may advantageously be removed from the beaker and placed on the litmus paper by means of a capillary tube.

Phosphorus.—The organic matter should be destroyed by one of the methods given in Bulletin 46, page 12, and phosphoric acid determined by either the gravimetric or volumetric molybdate method described in Bulletin 46, pages 12 and 14.

Chlorin.—Determine chlorin by titration with sulphocyanid, according to Volhard. For ordinary purposes the solution of the ash may be employed. More exact results may be obtained by dissolving about 1 gram of the meat extract in 20 cc of a 5 per cent solution of sodium carbonate, evaporating to dryness, and thoroughly igniting. The residue is then extracted with hot water, filtered and washed, after which the filter and contents are returned to a platinum dish and ignited. The contents of the dish are then dissolved in nitric acid, added to the filtrate, and the chlorin content determined as indicated above.

INSECTICIDES AND FUNGICIDES.

Seventeenth Convention, 1900, Bul. 62.

Methods suggested for trial, first year's work.

Eighteenth Convention, 1901, Bul. 67.

Methods I and II for the determination of total arsenious oxid were adopted as provisional:

TOTAL ARSENIOUS OXID.

Method I. [Made official in 1903.]

Solutions required:

Starch solution.—To prepare the starch solution, boil 2 grams of starch with 200 cc of water for about 5 minutes.

Iodin solution.—To prepare the standard iodin solution, dissolve 12.7 grams of powdered iodin in about 250 cc of water to which has been added 18 to 25 grams of c. p. potassium iodid, and make the whole up to a volume of 1 liter. [Made 2 liters in 1904.] To standardize this solution weigh out 1 gram of the inclosed dry c. p. arsenious oxid; transfer to a 250 cc flask by means of about 100 cc of a solution containing 2 grams of sodium hydrate in each 100 cc, and boil until all arsenious oxid goes in solution; cool; make to a volume of 250 cc and use 50 cc for analysis.

This 50 cc portion is concentrated by boiling in a 250 cc flask to half its volume and allowed to cool to 80° C. An equal volume of concentrated hydrochloric acid is now added accompanied by 3 grams of potassium iodid, mixed, and the whole allowed to stand for 10 minutes (to reduce the arsenic oxid, formed on boiling an alkaline arsenite, to arsenious oxid). The brown solution is then diluted with cold water and an approximately N/10 solution of sodium thiosulphate added, drop by drop, until the solution becomes exactly colorless. (This end point is easy to read without the aid of starch.) This solution is then made slightly alkaline with dry sodium carbonate (using a drop of methyl orange to read the change), and made slightly acid with hydrochloric acid, *taking care that all lumps of sodium carbonate on the bottom are acted on by the hydrochloric acid.* Sodium bicarbonate is now added in excess and the solution of iodin run in, drop by drop, using starch solution to read the end reaction. (Sometimes the solution gets dark toward the end of the titration. This must not be confused with the final dark blue color given by the iodin and starch.)

From the number of cubic centimeters of iodin solution used and the weight of arsenious oxid taken, the value of each cubic centimeter of iodin in arsenious oxid can be determined.

Method (Smith modified by Haywood).—Two grams of paris green are weighed out and transferred to a 250 cc flask and about 100 cc of water and 2 grams of sodium hydrate added. This solution is boiled for 5 to 10 minutes, or until all of the green particles have changed to red cuprous oxid. It is then cooled to room temperature and the volume made to 250 cc. The well-shaken liquid is filtered through a dry filter and 50 cc taken for analysis. The analysis is carried out from this point forward the same as when we standardize the iodin solution. (In duplicate.)

Method II. (Provisional.) a

Solutions.—The solutions required are the same as above, with the addition of a solution containing 2 to 3 grams of sodium potassium tartrate in 50 cc of water.

Method (Avery-Beans).—Sample the paris green (as one would an ore for assaying) down to about 1 gram. Pulverize this small sample in an agate mortar and weigh out 0.2 to 0.3 gram in a beaker of, say, 300 cc capacity. Add 25 cc of water, and to the green suspended in the water add, with constant stirring, concentrated hydrochloric acid till solution is just effected. Six drops are usually sufficient. Now add to the acid solution sodium carbonate solution till a slight permanent precipitate is formed. Dissolve this precipitate by adding 2 to 3 grams of sodium potassium tartrate in solution. Now dilute to about 200 cc, add solid sodium bicarbonate and starch solution, and titrate with iodin in the usual way. (In duplicate.)

Nineteenth Convention, 1902, Bul. 73.

Method I, provisional, for the determination of total arsenious oxid in paris green, was recommended for adoption as an official method. [Confirmed in 1903.]

Twenty-first Convention, 1904, Bul. 90, Cir. 20.

Use N/20 instead of N/10 iodin in titration of arsenic, i. e., Method I, iodin solution, line 3, for "1 liter" read "2 liters."

The electrolytic method was adopted as official for the determination of copper in paris green and copper carbonate.

This method is slightly modified as follows when used for this determination:

TOTAL COPPER OXID, METHOD I (OFFICIAL).

METHOD.

The cuprous oxid obtained in Method I for total arsenious oxid, by boiling the paris green with sodium hydroxid, is poured on the filter and well washed with hot water, after an aliquot portion of the filtrate has been taken for the determination of arsenious oxid. It is then dissolved in hot dilute nitric acid and made up to a volume of 250 cc. Fifty to 100 cc of this solution is used for the electrolytic determination of copper, as described on page 37, paragraph 2–b of Bulletin 46 (revised), Bureau of Chemistry, U. S. Department of Agriculture.

---

a This method should be dropped, as it has been adopted in modified form as an optional official method (1905).

The volumetric silver nitrate method for determining cyanogen in potassium cyanid was adopted as official (using N/20 instead of N/10 solution of silver nitrate):

POTASSIUM CYANID.

CYANOGEN (OFFICIAL).

SOLUTION REQUIRED.

A twentieth-normal solution of silver nitrate.

METHOD.

A large quantity of the sample is weighed out in a weighing bottle, dissolved in water, and made up to a definite volume. Aliquot portions of this are taken for analysis. The twentieth-normal silver nitrate solution is added a drop at a time with constant stirring, until one drop produces a permanent turbidity. In calculating the results, one equivalent of silver is equal to two equivalents of cyanogen, according to the following equation:

$$2 \; KCN + AgNO_3 = KCN.AgCN + KNO_3.$$

The Kissling method of determining nicotin was adopted as official: .

TOBACCO AND TOBACCO EXTRACTS.

NICOTIN, METHOD I (OFFICIAL).

SOLUTIONS REQUIRED.

An alcoholic soda solution containing 6 grams of sodium hydroxid, 40 cc of water, and 60 cc of 90 per cent alcohol. A weak sodium hydroxid solution containing 4 grams of sodium hydroxid in 1,000 cc of water. A standard sulphuric acid solution.

METHOD.

About 5 to 6 grams of tobacco extract or 20 grams of finely powdered tobacco, which has been previously dried at 60° C, so as to allow it to be powdered, is weighed out in a small beaker. Ten cubic centimeters of the alcoholic soda solution is added, followed, in the case of the tobacco extract, with enough C. P. powdered calcium carbonate to form a moist but not lumpy mass. The whole is well mixed. This is transferred to a Soxhlet extractor and exhausted for about 5 hours with ether. The ether is evaporated off at a low temperature by being held over the steam bath, and the residue is taken up with 50 cc of the weak soda solution mentioned above under "solutions required." This is transferred by means of water to a Kjeldahl flask, capable of holding about 500 cc, and distilled in a current of steam, using a condenser through which water is flowing rapidly. A three-bend outflow tube is used, and a few pieces of pumice and a small piece of paraffin are added to prevent bumping and frothing. The distillation is continued till all the nicotin has passed over, the distillate usually varying from 400 to 500 cc. When the distillation is complete only about 15 cc of the liquid should remain in the distillation flask. The distillate is titrated with standard sulphuric acid, using phenacetolin or cochineal as indicator. One molecule of sulphuric acid is equivalent to two molecules of nicotin.

The Avery hydrogen-peroxid method for determining sulphur in sulphur dips, etc., was adopted as provisional:

LIME-SULPHUR DIPS AND LIME-SULPHUR-SALT MIXTURE (PROVISIONAL).

TOTAL SULPHUR.

SOLUTIONS REQUIRED.

A saturated potassium hydroxid solution or a solution of caustic soda containing 100 grams to 100 cc of water. A 10 per cent barium chlorid solution. An approximately 3 per cent solution of hydrogen peroxid free from sulphates; if it contains sulphates add freshly precipitated barium carbonate and shake occasionally for several hours, then filter and use the clear solution.

METHOD.

Measure off 10 cc of the clear sample in a 100-cc measuring flask and fill to the mark. Use aliquot portions of 10 cc of this for analysis. Treat this aliquot with 3 cc of the caustic potash or soda solution, following this by 50 cc of hydrogen peroxid free of sulphates. Heat on the steam bath for just one-half hour and then acidify with hydrochloric acid, precipitate with barium chlorid in the ordinary manner in boiling solution, and finally weigh as barium sulphate.

Twenty-second Convention, 1905, Bul. 99, Cir. 26.

Methods I and IV, modifications of the Avery-Beans of determining total arsenic in paris green were adopted as optional official methods:

Method I. A weighed sample of the green was boiled with sodium acetate as in the determination of soluble arsenic by his (Avery's) method, the mixture filtered through asbestos, the soluble arsenic determined in the filtrate and the residue of the asbestos dissolved in dilute hydrochloric acid and titrated according to the Avery-Beans method (Method II, provisional).

Method IV. Take 0.4 gram of the finely ground green and boil with 25 cc of sodium acetate (containing 12 grams) for 10 minutes. Then add concentrated hydrochloric acid a drop at a time until solution is effected (about 10 cc of the acid will be necessary). Add concentrated sodium carbonate solution a drop at a time until a slight precipitate appears, then add a solution containing 2 to 3 grams of sodium potassium tartrate and finally sodium bicarbonate in excess. Titrate with iodin in the usual manner (Haywood's modification).

The thiosulphate method was adopted as an optional official method, using N/20 instead of N/10 thiosulphate solution.

TOTAL COPPER OXID (OPTIONAL OFFICIAL).

SOLUTIONS REQUIRED.

A standard twentieth-normal solution of sodium thiosulphate is prepared by dissolving 24.8 grams of the crystallized salt in 2 liters of water. This solution is standardized against C. P. copper foil dissolved in nitric acid by the method of analysis given in the following paragraph.

METHOD.

An aliquot portion of the nitric acid solution of copper oxid, used in Method I for total copper oxid, is made alkaline with sodium carbonate, then made slightly acid with acetic acid, diluted with water, and a considerable quantity of solid potassium iodid added, about 2 or 3 grams. When it is all dissolved by shaking, the free iodin is titrated with the twentieth-normal thiosulphate, using starch as indicator.

METHODS FOR THE ANALYSIS OF MEDICINAL PLANTS AND DRUGS.

Twenty-second Convention, 1905, Bul. 99, Cir. 26.

The method prescribed by the eighth revision of the United States Pharmacopœia, page 329, for the assay of gum opium, powdered opium, deodorized opium, and granulated opium, was adopted as a provisional method.

PROVISIONAL METHOD FOR THE ASSAY OF OPIUM (GUM, POWDERED, DEODORIZED, AND GRANULATED.)

[Eight Revision, U. S. Pharmacopœia, page 329.]

Assay of opium.

Opium, in any condition, to be valued.. 10.0 gm
Ammonia water.. 3.5 cc
Alcohol, ether, distilled water, lime water, each a sufficient quantity.

Introduce the opium (which, if fresh, should be in very small pieces, and if dry, in very fine powder) into an Erlenmeyer flask having a capacity of about 300 cc, add 100 cc of distilled water, stopper the flask, and agitate it every ten minutes (or continuously in a mechanical shaker) during three hours. Then pour the contents as evenly as possible upon a wetted filter having a diameter of 12 cm, and, when the liquid has drained off, wash the residue with distilled water, carefully dropped upon the edges of the filter and its contents, until 150 cc of filtrate have been obtained. Then carefully transfer the moist opium back into the flask by means of a spatula, add 50 cc of distilled water, agitate it thoroughly and repeatedly during fifteen minutes, and return the whole to the filter. When the liquid has drained off, wash the residue, as before, until the second filtrate measures 150 cc, and finally collect about 20 cc more of a third filtrate. Evaporate carefully in a tared dish, first, the second filtrate to a small volume, then add the first filtrate, rinsing the vessels with the third filtrate, and continue the evaporation until the residue weighs 14 gm. Rotate the concentrated solution about in the dish until the rings of extract are redissolved, pour the liquid into a tared Erlenmeyer flask having a capacity of about 100 cc, and rinse the dish with a few drops of water at a time until the entire solution, after the rinsings have been added to the flask, weighs 20 gm. Then add 10 gm (or 12.2 cc) of alcohol, shake the flask well, add 25 cc of ether, and repeat the shaking. Now add the ammonia water from a graduated pipette or burette, stopper the flask with a sound cork, shake it thoroughly during ten minutes, and then set it aside in a moderately cool place for at least six hours, or over night.

Remove the stopper carefully and should any crystals adhere to it brush them into the flask. Place in a small funnel two rapidly acting filters of a diameter of 7 cm, plainly folded one within the other (the triple fold of the inner filter being laid against the single side of the outer filter), wet them well with ether, and decant the ethereal solution as completely as possible upon the inner filter.

Add 10 cc of ether to the contents of the flask, rotate it, and again decant the ethereal layer upon the inner filter. Repeat this operation with another portion of 10 cc of ether. Then pour the liquid in the flask into the filter in portions in such a way as to transfer the greater portion of the crystals to the filter, and, when the liquid has passed through, transfer the remaining crystals to the filter by washing the flask with several portions of water, using not more than 15 cc in all. Use a feather or rubber-tipped glass rod to remove the crystals that adhere to the flask. Allow the double filter to drain, then apply water to the crystals, drop by drop, until they are practically free from mother liquor, and afterwards wash them, drop by drop, from a pipette, with alcohol previously saturated with powdered morphine. When this has passed through, displace the remaining alcohol by ether, using about 10 cc or more if necessary. Allow the filter to dry in a moderately warm place, at a temperature not exceeding 60° C. (140° F.) until its weight remains constant, then carefully transfer the crystals to a tared watch-glass and weigh them.

Place the crystals (which are not quite pure) in an Erlenmeyer flask, add lime water (10 cc for each 0.1 gm of morphine) and shake the flask at intervals during half an hour. Pass the liquid through two counterpoised, rapidly acting, plainly folded filters, one within the other (the triple fold of the inner filter being laid against the single fold of the outer filter), rinse the flask with more lime water, and pass the washings through the filter until the filtrate, after acidulating, no longer yields a precipitate with mercuric potassium iodide T. S. Press the filters until nearly dry between bibulous paper and dry them to a constant weight; then weigh the contents, using the outer filter as a counterpoise. Deduct the weight of the insoluble matter on the filter from the weight of the impure morphine previously found. The difference multiplied by 10 represents the percentage of crystallized morphine contained in the opium.

O

CPSIA information can be obtained
at www.ICGtesting.com
Printed in the USA
BVHW07s1306280918
528774BV00021B/1423/P